PERIMETER

THINGS YOU SHOULD KNOW
(QUESTIONS AND ANSWERS)

By Rumi Michael Leigh

Introduction

I would like to thank you for purchasing this book, *"Perimeter, things you should know (questions and answers)"*.

This book will help you understand, revise, and have a good general knowledge and understanding of the basics of perimeter exercises.

I hope you enjoy it!

Table of Contents

Part 1: Perimeter

Exercise 1

Questions

a) A rectangle has a length of 12 cm and a width of 8 cm. What is its perimeter?
b) A square has a side length of 6 cm. What is its perimeter?
c) A regular hexagon has a side length of 5 cm. What is its perimeter?
d) A circle has a radius of 8 cm. What is its circumference (perimeter)?
e) A triangle has sides of length 6 cm, 8 cm, and 10 cm. What is its perimeter?

Answers

a) A rectangle has a length of 12 cm and a width of 8 cm. What is its perimeter?

$P = 2l + 2w$

$P = 2(12) + 2(8) = 24 + 16 = 40$ cm

Answer: the perimeter of the rectangle is 40 cm.

b) A square has a side length of 6 cm. What is its perimeter?

$P = 4s$

$P = 4(6) = 24$ cm

Answer: the perimeter of the square is 24 cm.

c) A regular hexagon has a side length of 5 cm. What is its perimeter?

$P = 6s$

$P = 6(5) = 30$ cm

Answer: the perimeter of the regular hexagon is 30 cm.

d) A circle has a radius of 8 cm. What is its circumference (perimeter)?

$C = 2\pi r$

$\pi = 3.14$

$C = 2(3.14)(8) = 50.24$ cm

Answer: the circumference (perimeter) of the circle is 50.24 cm.

e) A triangle has sides of length 6 cm, 8 cm, and 10 cm. What is its perimeter?

$P = 6 + 8 + 10 = 24$ cm

Answer: the perimeter of the triangle is 24 cm.

Exercise 2

Questions

a) A parallelogram has a base of 10 cm and a height of 6 cm. What is its perimeter?

b) A trapezoid has parallel sides of lengths 6 cm and 12 cm, and a height of 4 cm. What is its perimeter?

c) A regular octagon has a side length of 7 cm. What is its perimeter?

d) A kite has diagonals of lengths 10 cm and 8 cm. What is its perimeter?

e) A rectangle has a perimeter of 30 cm and a length of 8 cm. What is its width?

Answers

a) A parallelogram has a base of 10 cm and a height of 6 cm. What is its perimeter?

$P = 2(b + h)$

$P = 2(10 + 6) = 32$ cm

Answer: the perimeter of the parallelogram is 32 cm.

b) A trapezoid has parallel sides of lengths 6 cm and 12 cm, and a height of 4 cm. What is its perimeter?

To find the perimeter of a trapezoid, we need to know the length of its non-parallel sides. However, this information is not given in the problem, so we cannot calculate the perimeter.

c) A regular octagon has a side length of 7 cm. What is its perimeter?

$P = 8s$

$P = 8(7) = 56$ cm

Answer: the perimeter of the regular octagon is 56 cm.

d) A kite has diagonals of lengths 10 cm and 8 cm. What is its perimeter?

To find the perimeter of a kite, we need to know the lengths of its four sides. However, this information is not given in the problem, so we cannot calculate the perimeter.

e) A rectangle has a perimeter of 30 cm and a length of 8 cm. What is its width?

P = 2l + 2w

30 = 2(8) + 2w

w = (30 - 2(8))/2 = 7 cm

Answer: the width of the rectangle is 7 cm.

Exercise 3

Questions

a) A regular pentagon has a perimeter of 25 cm. What is the length of each side?

b) A circle has a diameter of 10 cm. What is its circumference (perimeter)?

c) A regular hexagon has a perimeter of 24 cm. What is the length of each side?

d) A square has a perimeter of 28 cm. What is the length of each side? Solution:

e) A regular decagon has a perimeter of 50 cm. What is the length of each side?

Answers

a) A regular pentagon has a perimeter of 25 cm. What is the length of each side?

P = 5s

s = P/5 = 25/5 = 5 cm

Answer: the length of each side of the regular pentagon is 5 cm.

b) A circle has a diameter of 10 cm. What is its circumference (perimeter)?

C = πd

C = 3.14(10) = 31.4 cm

Answer: the circumference (perimeter) of the circle is 31.4 cm.

c) A regular hexagon has a perimeter of 24 cm. What is the length of each side?

P = 6s

s = P/6 = 24/6 = 4 cm

Answer: the length of each side of the regular hexagon is 4 cm.

d) A square has a perimeter of 28 cm. What is the length of each side? Solution:

P = 4s

s = P/4 = 28/4 = 7 cm

Answer: the length of each side of the square is 7 cm.

e) A regular decagon has a perimeter of 50 cm. What is the length of each side?

P = 10s

s = P/10 = 50/10 = 5 cm

Answer: the length of each side of the regular decagon is 5 cm.

Exercise 4

Questions

a) A rectangle has a length of 12 cm and a width of 8 cm. What is its perimeter?

b) A circle has a diameter of 10 cm. What is its circumference?

c) A regular octagon has a side length of 4 cm. What is its perimeter?

d) A regular hexagon has a perimeter of 30 cm. What is its side length?

Answers

a) A rectangle has a length of 12 cm and a width of 8 cm. What is its perimeter?

P = 2l + 2w

P = 2(12) + 2(8)

P = 24 + 16

P = 40 cm

Answer: the perimeter of the rectangle is 40 cm.

b) A circle has a diameter of 10 cm. What is its circumference?

C = πd

C = π(10)

C ≈ 31.42 cm

Answer: the circumference of the circle is approximately 31.42 cm.

c) A regular octagon has a side length of 4 cm. What is its perimeter?

P = 8s

P = 8(4)

P = 32 cm

Answer: the perimeter of the octagon is 32 cm.

d) A regular hexagon has a perimeter of 30 cm. What is its side length?

s = P/6

s = 30/6

s = 5 cm

Answer: the side length of the hexagon is 5 cm.

Exercise 5

Questions

a) A regular heptagon has a perimeter of 35 cm. What is its side length?
b) A parallelogram has a base of length 12 cm and a height of 8 cm. If one of its sides is 10 cm long, what is its perimeter?
c) A regular octagon has a side length of 5 cm. What is its perimeter?
d) A rectangle has a perimeter of 32 cm and a length of 10 cm. What is its width?
e) A square has a perimeter of 20 cm. What is the length of each side?

Answers

a) A regular heptagon has a perimeter of 35 cm. What is its side length?

s = P/7

s = 35/7

s = 5 cm

Answer: the side length of the heptagon is 5 cm.

b) A parallelogram has a base of length 12 cm and a height of 8 cm. If one of its sides is 10 cm long, what is its perimeter?

Since opposite sides of a parallelogram are congruent, the other side of the parallelogram is also 10 cm long.

The perimeter can be found by adding up the lengths of all four sides.

Perimeter = 10 + 10 + 12 + 12

Perimeter = 44 cm

Answer: the perimeter of the parallelogram is 44 cm.

c) A regular octagon has a side length of 5 cm. What is its perimeter?

P = 8s

P = 8(5)

P = 40 cm

Answer: the perimeter of the octagon is 40 cm.

d) A rectangle has a perimeter of 32 cm and a length of 10 cm. What is its width?

Perimeter = 2(length + width)

32 = 2(10 + w)

16 = 10 + w

w = 6 cm

Answer: the width of the rectangle is 6 cm.

e) A square has a perimeter of 20 cm. What is the length of each side?

P = 4s, where s is the length of each side.

20 = 4s

s = 5 cm

Answer: the length of each side of the square is 5 cm.

Part 2: Perimeter

Exercise 1

Questions

a) A triangle has sides of length 5 cm, 7 cm, and 8 cm. What is its perimeter?

b) A rectangle has a length of 6 cm and a width of 4 cm. What is its perimeter?

c) A regular hexagon has a side length of 3 cm. What is its perimeter?

d) A parallelogram has a base of length 10 cm and a height of 6 cm. What is its perimeter if its other side is also 10 cm long?

Answers

a) A triangle has sides of length 5 cm, 7 cm, and 8 cm. What is its perimeter?

The perimeter of the triangle is the sum of the lengths of its three sides.

Perimeter = 5 + 7 + 8

Perimeter = 20 cm

Answer: the perimeter of the triangle is 20 cm.

b) A rectangle has a length of 6 cm and a width of 4 cm. What is its perimeter?

P = 2(length + width)

P = 2(6 + 4)

P = 20 cm

Answer: the perimeter of the rectangle is 20 cm.

c) A regular hexagon has a side length of 3 cm. What is its perimeter?

P = 6s, where s is the side length.

P = 6(3)

P = 18 cm

Answer: the perimeter of the hexagon is 18 cm.

d) A parallelogram has a base of length 10 cm and a height of 6 cm. What is its perimeter if its other side is also 10 cm long?

Solution: Since opposite sides of a parallelogram are congruent, the length of the other side is also 10 cm.

The perimeter can be found by adding up the lengths of all four sides.

Perimeter = 10 + 10 + 6 + 6

Perimeter = 32 cm

Answer: the perimeter of the parallelogram is 32 cm.

Exercise 2

Questions

a) A regular pentagon has a perimeter of 30 cm. What is the length of each side?

b) A triangle has a perimeter of 15 cm. If two of its sides have lengths 4 cm and 6 cm, what is the length of the third side?

c) A square has an area of 16 cm^2. What is its perimeter?

d) A rectangle has a perimeter of 18 cm and a length of 5 cm. What is its width?

   Perimeter = 2(length + width)

Answers

a) A regular pentagon has a perimeter of 30 cm. What is the length of each side?

   s = P/5

   s = 30/5

   s = 6 cm

   Answer: the length of each side of the pentagon is 6 cm.

b) A triangle has a perimeter of 15 cm. If two of its sides have lengths 4 cm and 6 cm, what is the length of the third side?

   Let x be the length of the third side.

   Perimeter = 4 + 6 + x

   15 = 10 + x

   x = 5 cm

   Answer: the length of the third side is 5 cm.

c) A square has an area of 16 cm^2. What is its perimeter?

   A = s^2, where s is the length of each side.

   s = sqrt(A)

   s = sqrt(16)

   s = 4 cm

P = 4s

P = 4(4)

P = 16 cm

Answer: the perimeter of the square is 16 cm.

d) A rectangle has a perimeter of 18 cm and a length of 5 cm. What is its width?

Perimeter = 2(length + width)

18 = 2(5 + w)

9 = 5 + w

w = 4 cm

Answer: the width of the rectangle is 4 cm.

Exercise 3

Questions

a) A regular hexagon has a perimeter of 36 cm. What is the length of each side?

b) A regular octagon has a perimeter of 32 cm. What is the length of each side?

c) A rectangle has a perimeter of 42 cm and an area of 72 cm^2. What are its dimensions?

d) A trapezoid has a height of 8 cm and bases of lengths 5 cm and 9 cm. What is its perimeter if its non-parallel sides have lengths of 6 cm and 7 cm?

Answers

a) A regular hexagon has a perimeter of 36 cm. What is the length of each side?

s = P/6

s = 36/6

s = 6 cm

Answer: the length of each side of the hexagon is 6 cm.

b) A regular octagon has a perimeter of 32 cm. What is the length of each side?

s = P/8, where P is the perimeter.

s = 32/8

s = 4 cm

Answer: the length of each side of the octagon is 4 cm.

c) A rectangle has a perimeter of 42 cm and an area of 72 cm^2. What are its dimensions?

Perimeter = 2(L + W)

42 = 2(L + W)

L + W = 21

A = LW

72 = LW

L = 72/W

72/W + W = 21

Multiplying both sides by W, we get: 72 + W^2 = 21W

Rearranging this equation, we get: W^2 - 21W + 72 = 0

We can solve this quadratic equation using factoring or the quadratic formula.

Factoring, we get: (W - 3)(W - 18) = 0

This gives us two possible values for W: W = 3 or W = 18. If we substitute each value back into the equation for L, we get: L = 72/3 = 24 or L = 72/18 = 4

Answers: the dimensions of the rectangle are either 24 cm by 3 cm or 18 cm by 4 cm.

d) A trapezoid has a height of 8 cm and bases of lengths 5 cm and 9 cm. What is its perimeter if its non-parallel sides have lengths of 6 cm and 7 cm?

To find the perimeter of the trapezoid, we need to first find the lengths of the parallel sides.

Area = (b1 + b2)h/2

8 = (5 + 9)s/2

16 = 14s/2

s = 16/7 cm

The lengths of the parallel sides are 5 cm and 9 cm, and the lengths of the non-parallel sides are 6 cm and 7 cm.

Perimeter = 5 + 6 + 7 + 9 + 16/7 + 16/7

Perimeter ≈ 34.29 cm

Answer: the perimeter of the trapezoid is approximately 34.29 cm.

Exercise 4

Questions

a) A regular decagon has a perimeter of 50 cm. What is the length of each side?

b) A regular pentagon has a perimeter of 25 cm. What is the length of each side?

c) A regular hexagon has a perimeter of 36 cm. What is the length of each side?

d) A rectangle has a length of 8 cm and a width of 4 cm. What is its perimeter?

Answers

a) A regular decagon has a perimeter of 50 cm. What is the length of each side?

s = P/10

s = 50/10

s = 5 cm

Answer: the length of each side of the decagon is 5 cm.

b) A regular pentagon has a perimeter of 25 cm. What is the length of each side?

s = P/5

s = 25/5

s = 5 cm

Answer: the length of each side of the pentagon is 5 cm.

c) A regular hexagon has a perimeter of 36 cm. What is the length of each side?

s = P/6

s = 36/6

s = 6 cm

Answer: the length of each side of the hexagon is 6 cm.

d) A rectangle has a length of 8 cm and a width of 4 cm. What is its perimeter?

P = 2L + 2W

P = 2(8) + 2(4)

P = 16 + 8

P = 24 cm

Answer: the perimeter of the rectangle is 24 cm.

Exercise 5

Questions

a) A square has a perimeter of 20 cm. What is the length of each side?

b) A regular pentagon has a perimeter of 25 cm. What is the length of each side?

c) A triangle has sides of lengths 3 cm, 4 cm, and 5 cm. What is its perimeter?

d) A parallelogram has a base of 8 cm and a height of 4 cm. What is its perimeter?

e) A regular hexagon has a perimeter of 36 cm. What is the length of each side?

Answers

a) A square has a perimeter of 20 cm. What is the length of each side?

$s = P/4$

$s = 20/4$ $s = 5$ cm

Answer: the length of each side of the square is 5 cm.

b) A regular pentagon has a perimeter of 25 cm. What is the length of each side?

$s = P/5$

$s = 25/5$

$s = 5$ cm

Answer: the length of each side of the pentagon is 5 cm.

c) A triangle has sides of lengths 3 cm, 4 cm, and 5 cm. What is its perimeter?

$P = 3 + 4 + 5$

$P = 12$ cm

Answer: the perimeter of the triangle is 12 cm.

d) A parallelogram has a base of 8 cm and a height of 4 cm. What is its perimeter?

$P = 2b + 2h$

$P = 2(8) + 2(4)$

$P = 16 + 8$

$P = 24$ cm

Answer: the perimeter of the parallelogram is 24 cm.

e) A regular hexagon has a perimeter of 36 cm. What is the length of each side?

$s = P/6$

$s = 36/6$

$s = 6$ cm

Answer: the length of each side of the hexagon is 6 cm.

Part 3: Perimeter

Exercise 1

Questions

    a) A rectangle has a length of 12 cm and a width of 5 cm. What is its perimeter?

    b) A regular octagon has a perimeter of 48 cm. What is the length of each side?

    c) A square has an area of 49 cm^2. What is its perimeter?

    d) A triangle has sides of lengths 3 cm, 4 cm, and 5 cm. What is its perimeter?

    e) A regular heptagon has a side length of 6 cm. What is its perimeter?

Answers

    a) A rectangle has a length of 12 cm and a width of 5 cm. What is its perimeter?

       Perimeter = 2(length + width)

       P = 2(12 cm + 5 cm)

       P = 2(17 cm) = 34 cm

       Answer: the perimeter of the rectangle is 34 cm.

    b) A regular octagon has a perimeter of 48 cm. What is the length of each side?

       Perimeter of octagon = 8(side length) = 48 cm

       Side length = 48 cm / 8 = 6 cm

       Answer: the length of each side is 6 cm.

    c) A square has an area of 49 cm^2. What is its perimeter?

       Area of square = side^2 = 49 cm^2

       Side length = sqrt(49 cm^2) = 7 cm

       Perimeter = 4(side length) = 4(7 cm) = 28 cm

       Answer: the perimeter of the square is 28 cm.

    d) A triangle has sides of lengths 3 cm, 4 cm, and 5 cm. What is its perimeter?

       Perimeter of triangle = 3 cm + 4 cm + 5 cm = 12 cm

       Answer: the perimeter of the triangle is 12 cm

    e) A regular heptagon has a side length of 6 cm. What is its perimeter?

       Perimeter of heptagon = 7(side length) = 7(6 cm) = 42 cm

       Answer: the perimeter of the heptagon is 42 cm

Exercise 2

Questions

a) A trapezoid has a height of 8 cm, and the lengths of its two bases are 6 cm and 12 cm. What is its perimeter?

b) A circle has a radius of 5 cm. What is its circumference?

c) A regular hexagon has an apothem (the distance from the center to a side) of 4 cm. What is its perimeter?

d) A rectangle has a perimeter of 36 cm and a length of 12 cm. What is its width?

Answers

a) A trapezoid has a height of 8 cm, and the lengths of its two bases are 6 cm and 12 cm. What is its perimeter?

Perimeter of trapezoid = 8 cm + 6 cm + 12 cm + 10 cm = 36 cm

Answer: the perimeter of the trapezoid is 36 cm.

b) A circle has a radius of 5 cm. What is its circumference?

Circumference of circle = 2π(radius) = 2π(5 cm) = 10π cm

Answer: the circumference of the circle is 10π cm

c) A regular hexagon has an apothem (the distance from the center to a side) of 4 cm. What is its perimeter?

A regular hexagon can be divided into 6 equilateral triangles.

The apothem of each equilateral triangle = 4 cm

The side length of each equilateral triangle = 2(apothem) = 8 cm

Perimeter of hexagon = 6(side length) = 6(8 cm) = 48 cm

Answer: the perimeter of the hexagon is 48 cm

d) A rectangle has a perimeter of 36 cm and a length of 12 cm. What is its width?

Perimeter of rectangle = 2(length + width) = 36 cm

Length = 12 cm

Width = (perimeter - 2(length)) / 2 = (36 cm - 2(12 cm)) / 2 = 6 cm

Answer: the width of the perimeter is 6 cm

Exercise 3

Questions

a) A regular pentagon has a perimeter of 35 cm. What is the length of each side?
b) A circle has a circumference of 12π cm. What is its radius?
c) A regular decagon has a side length of 7 cm. What is its perimeter?

Answers

a) A regular pentagon has a perimeter of 35 cm. What is the length of each side?

Perimeter of pentagon = 5(side length) = 35 cm

Side length = 35 cm / 5 = 7 cm

Answer: the length of each side is 7 cm

b) A circle has a circumference of 12π cm. What is its radius?

Circumference of circle = 2π(radius) = 12π cm

Radius = (circumference) / (2π) = (12π cm) / (2π) = 6 cm

Answer: the radius is 6 cm

c) A regular decagon has a side length of 7 cm. What is its perimeter?

Perimeter of decagon = 10(side length) = 10(7 cm) = 70 cm

Answer: the perimeter is 70 cm

Exercise 4

Questions

a) Find the perimeter of a square with a side length of 10 cm.
b) Find the perimeter of a rectangle with length 15 cm and width 8 cm.
c) Find the perimeter of a regular pentagon with a side length of 6 cm.
d) The perimeter of a rectangle is 40 cm, and its width is 4 cm. Find its length.
e) The perimeter of a regular hexagon is 36 cm. Find the length of one side.

Answers

a) Find the perimeter of a square with a side length of 10 cm.

Perimeter of square = 4(side length) = 4(10 cm) = 40 cm

Answer: the perimeter of the square is 40 cm

b) Find the perimeter of a rectangle with length 15 cm and width 8 cm.

Perimeter of rectangle = 2(length + width) = 2(15 cm + 8 cm) = 46 cm

Answer: the perimeter of the rectangle is 46 cm

c) Find the perimeter of a regular pentagon with a side length of 6 cm.

Perimeter of regular pentagon = 5(side length) = 5(6 cm) = 30 cm

Answer: the perimeter of the pentagon is 30 cm

d) The perimeter of a rectangle is 40 cm, and its width is 4 cm. Find its length.

Perimeter of rectangle = 2(length + width) = 40 cm Width = 4 cm

Length + 4 cm + Length + 4 cm = 40 cm

2Length + 8 cm = 40 cm

2Length = 32 cm

Length = 16 cm

Answer: the length of the rectangle is 16 cm

e) The perimeter of a regular hexagon is 36 cm. Find the length of one side.

Perimeter of regular hexagon = 6(side length) = 36 cm

Side length = 36 cm / 6 = 6 cm

Answer: the length of one side is 6 cm

Exercise 5

Questions

a) The perimeter of a parallelogram is 28 cm. If its base is 7 cm, find its height.
b) Find the perimeter of a trapezium with parallel sides of length 5 cm and 12 cm, and a height of 8 cm.
c) A regular polygon with 10 sides has a perimeter of 60 cm. Find the length of one side.

Answers

a) The perimeter of a parallelogram is 28 cm. If its base is 7 cm, find its height.

Perimeter of parallelogram = 2(base + height) = 28 cm

Base = 7 cm

2(7 cm + height) = 28 cm

7 cm + height = 14 cm

Height = 7 cm

Answer: the height of the parallelogram is 28 cm

b) Find the perimeter of a trapezium with parallel sides of length 5 cm and 12 cm, and a height of 8 cm.

Perimeter of trapezium = sum of all sides = 5 cm + 12 cm + 8 cm + 8 cm = 33 cm

Answer: the perimeter of the trapezium is 33 cm

c) A regular polygon with 10 sides has a perimeter of 60 cm. Find the length of one side.

Perimeter of 10-sided regular polygon = 60 cm

Length of one side = 60 cm / 10 = 6 cm

Answer: the length of one side is 6 cm

Part 4: Perimeter

Exercise 1

Questions

a) A regular polygon with 12 sides has a perimeter of 96 cm. Find the length of one side.
b) The perimeter of a rectangle is 38 cm, and its length is 11 cm. Find its width.
c) Find the perimeter of a trapezium with a height of 6 cm, parallel sides of length 8 cm and 12 cm, and a non-parallel side of length 7 cm.
d) Find the perimeter of an equilateral triangle with a side length of 5 cm.

Answers

a) A regular polygon with 12 sides has a perimeter of 96 cm. Find the length of one side.

Perimeter of 12-sided regular polygon = 96 cm

Length of one side = 96 cm / 12 = 8 cm

Answer: the length of one side of the polygon is 8 cm

b) The perimeter of a rectangle is 38 cm, and its length is 11 cm. Find its width.

Perimeter of rectangle = 2(length + width) = 38 cm

Length = 11 cm

2(11 cm + width) = 38 cm

22 cm + 2width = 38 cm

2width = 16 cm

Width = 8 cm

Answer: the width of the rectangle is 8 cm

c) Find the perimeter of a trapezium with a height of 6 cm, parallel sides of length 8 cm and 12 cm, and a non-parallel side of length 7 cm.

Perimeter of trapezium = sum of all sides = 8 cm + 12 cm + 7 cm + 7 cm + 6 cm + 6 cm = 46 cm

Answer: the perimeter of the trapezium is 46 cm

d) Find the perimeter of an equilateral triangle with a side length of 5 cm.

Perimeter of equilateral triangle = 3(side length) = 3(5 cm) = 15 cm

Answer: the perimeter of the triangle is 15 cm

Exercise 2

Questions

a) The perimeter of an isosceles triangle is 24 cm. If the two equal sides are each 7 cm long, what is the length of the third side?

b) The perimeter of a regular octagon is 48 cm. Find the length of each side of the octagon.

c) The perimeter of a rectangle is 42 cm. If the length of the rectangle is 3 times its width, what are the dimensions of the rectangle?

d) The perimeter of an equilateral triangle is 27 cm. What is the length of each side of the triangle?

e) The perimeter of a regular hexagon is 36 cm. Find the length of each side of the hexagon.

Answers

a) The perimeter of an isosceles triangle is 24 cm. If the two equal sides are each 7 cm long, what is the length of the third side?

Perimeter of isosceles triangle = a + b + c = 24 cm

Two equal sides = a = b = 7 cm

7 cm + 7 cm + c = 24 cm

c = 10 cm

Answer: the length of the third side is 10 cm

b) The perimeter of a regular octagon is 48 cm. Find the length of each side of the octagon.

Perimeter of regular octagon = 8(side length) = 48 cm

Side length = 6 cm

Answer: the length of each side of the octagon

c) The perimeter of a rectangle is 42 cm. If the length of the rectangle is 3 times its width, what are the dimensions of the rectangle?

Perimeter of rectangle = 2(length + width) = 42 cm

Let the width be x.

Length is 3x

2(3x + x) = 42 cm

8x = 42 cm

x = 5.25 cm (width)

length = 3x = 15.75 cm

Answers: the length is 15.75 cm and the width is 5.25 cm

d) The perimeter of an equilateral triangle is 27 cm. What is the length of each side of the triangle?

Perimeter of equilateral triangle = 3(side length) = 27 cm

Side length = 9 cm

Answer: the length of each side of the triangle is 9 cm

e) The perimeter of a regular hexagon is 36 cm. Find the length of each side of the hexagon.

Perimeter of regular hexagon = 6(side length) = 36 cm

Side length = 6 cm

Answer: the length of each side of the hexagon is 6 cm

Exercise 3

Questions

a) The perimeter of a parallelogram is 30 cm. If the height of the parallelogram is 4 cm, what is the length of the base?

b) The perimeter of a regular pentagon is 25 cm. Find the length of each side of the pentagon.

c) The perimeter of a square is 84 cm. Find the length of the diagonal of the square.

d) The perimeter of a rectangle is 60 cm. If the length of the rectangle is twice its width, what are the dimensions of the rectangle?

Answers

a) The perimeter of a parallelogram is 30 cm. If the height of the parallelogram is 4 cm, what is the length of the base?

Perimeter of parallelogram = 2(base + height) = 30 cm

2(base + 4 cm) = 30 cm

base + 4 cm = 15 cm

base = 11 cm

Answer: the length of the base is 11 cm

b) The perimeter of a regular pentagon is 25 cm. Find the length of each side of the pentagon.

Perimeter of regular pentagon = 5(side length) = 25 cm

Side length = 5 cm

Answer: the length of each side of the pentagon is 5cm

c) The perimeter of a square is 84 cm. Find the length of the diagonal of the square.

Perimeter of square = 4(side length) = 84 cm

Side length = 21 cm

Diagonal of square = side length x $\sqrt{2}$ = 21 cm x $\sqrt{2}$ = 29.7 cm

Answer: the length of the diagonal of the square is 29.7 cm

d) The perimeter of a rectangle is 60 cm. If the length of the rectangle is twice its width, what are the dimensions of the rectangle?

Perimeter of rectangle = 2(length + width) = 60 cm

Let the width be x.

Then the length is 2x. 2(2x + x) = 60 cm

6x = 60 cm

x = 10 cm (width)

length = 2x = 20 cm

Answers: the length is 20 cm and the width is 10 cm

Exercise 4

Questions

a) The perimeter of a rhombus is 40 cm. If one diagonal of the rhombus is 10 cm, what is the length of the other diagonal?

b) The perimeter of a regular heptagon is 35 cm. Find the length of each side of the heptagon.

c) The perimeter of an isosceles triangle is 40 cm. If the base is 12 cm, what is the length of each of the equal sides?

Answers

a) The perimeter of a rhombus is 40 cm. If one diagonal of the rhombus is 10 cm, what is the length of the other diagonal?

Perimeter of rhombus = 4(side length) = 40 cm

Side length = 10 cm

Formula for diagonals of a rhombus: $d1^2 + d2^2 = 4$(side length)^2

$10^2 + d2^2 = 4(10$ cm$)^2$

$d2^2 = 300$

$d2 = \sqrt{300} = 10\sqrt{3}$ cm

Answer: the length of the other diagonal is $10\sqrt{3}$ cm

b) The perimeter of a regular heptagon is 35 cm. Find the length of each side of the heptagon.

Perimeter of regular heptagon = 7(side length) = 35 cm

Side length = 5 cm

Answer: the length of each side of the heptagon is 5 cm

c) The perimeter of an isosceles triangle is 40 cm. If the base is 12 cm, what is the length of each of the equal sides?

Perimeter of isosceles triangle = 2(equal sides) + base = 40 cm

Base = 12 cm

2(equal sides) = 40 cm - 12 cm = 28 cm equal sides = 14 cm

Answer: the length of each of the equal sides is 14 cm

Part 5: Perimeter

Exercise 1

Questions

a) The perimeter of a regular pentagon is 25 cm. Find the length of each side of the pentagon.

b) The perimeter of an equilateral triangle is 27 cm. Find the length of each side of the triangle.

c) The perimeter of a regular hexagon is 36 cm. Find the length of each side of the hexagon.

d) The perimeter of a regular octagon is 32 cm. Find the length of each side of the octagon.

e) The perimeter of a rectangle is 34 cm. If the length is 2 cm more than twice the width, what are the dimensions of the rectangle?

Answers

a) The perimeter of a regular pentagon is 25 cm. Find the length of each side of the pentagon.

Perimeter of regular pentagon = 5(side length) = 25 cm

Side length = 5 cm

Answer: the length of each side of the pentagon is 5 cm

b) The perimeter of an equilateral triangle is 27 cm. Find the length of each side of the triangle.

Perimeter of equilateral triangle = 3(side length) = 27 cm

Side length = 9 cm

Answer: the length of each side of the triangle is 9 cm

c) The perimeter of a regular hexagon is 36 cm. Find the length of each side of the hexagon.

Perimeter of regular hexagon = 6(side length) = 36 cm

Side length = 6 cm

Answer: the length of each side of the hexagon is 6 cm

d) The perimeter of a regular octagon is 32 cm. Find the length of each side of the octagon.

Perimeter of regular octagon = 8(side length) = 32 cm

Side length = 4 cm

Answer: the length of each side of the octagon is 4 cm

e) The perimeter of a rectangle is 34 cm. If the length is 2 cm more than twice the width, what are the dimensions of the rectangle?

Perimeter of rectangle = 2(length + width) = 34 cm

length = 2 cm + 2(width)

2(2 cm + 2(width) + width) = 34 cm

6(width) = 30 cm width = 5 cm

length = 2 cm + 2(5 cm) = 12 cm

Answers: The dimensions of the rectangle are 12 cm x 5 cm.

Exercise 2

Questions

a) The perimeter of a square is 44 cm. What is the length of each side of the square?

b) The perimeter of a regular heptagon is 42 cm. Find the length of each side of the heptagon.

c) The perimeter of a parallelogram is 38 cm. If the height is 5 cm and one of the sides is 10 cm, what is the length of the other side?

d) The perimeter of an isosceles triangle is 36 cm. If the base is 14 cm, what is the length of each of the equal sides?

e) The perimeter of a regular nonagon is 45 cm. Find the length of each side of the nonagon.

Answers

a) The perimeter of a square is 44 cm. What is the length of each side of the square?

Perimeter of square = 4(side length) = 44 cm

Side length = 11 cm

Answer: the length of each side of the square is 11 cm

b) The perimeter of a regular heptagon is 42 cm. Find the length of each side of the heptagon.

Perimeter of regular heptagon = 7(side length) = 42 cm

Side length = 6 cm

Answer: the length of each side of the heptagon is 6 cm

c) The perimeter of a parallelogram is 38 cm. If the height is 5 cm and one of the sides is 10 cm, what is the length of the other side?

Perimeter of parallelogram = 2(base + side) = 38 cm

Height = 5 cm

Side = 10 cm

2(base + 10 cm) = 38 cm

Base = 9 cm

Answer: the length of the other side is 9 cm.

d) The perimeter of an isosceles triangle is 36 cm. If the base is 14 cm, what is the length of each of the equal sides?

Perimeter of isosceles triangle = 2(equal sides) + base = 36 cm

Base = 14 cm 2(equal sides) = 36 cm - 14 cm = 22 cm

equal sides = 11 cm

Answer: the length of each of the equal sides is 11 cm

e) The perimeter of a regular nonagon is 45 cm. Find the length of each side of the nonagon.

Perimeter of regular nonagon = 9(side length) = 45 cm

Side length = 5 cm

Answer: the length of each side of the nonagon is 5 cm

Exercise 3

Questions

a) The perimeter of a rectangle is 36 cm. If the length is 2 cm more than three times the width, what are the dimensions of the rectangle?

b) The perimeter of a regular decagon is 50 cm. Find the length of each side of the decagon.

c) The perimeter of an equilateral triangle is 36 cm. Find the length of each side of the triangle.

d) The perimeter of a square is 68 cm. What is the length of each side of the square?

Answers

a) The perimeter of a rectangle is 36 cm. If the length is 2 cm more than three times the width, what are the dimensions of the rectangle?

Perimeter of rectangle = 2(length + width) = 36 cm length = 2 cm + 3(width)

2(2 cm + 3(width) + width) = 36 cm

8(width) = 32 cm

width = 4 cm

length = 2 cm + 3(4 cm) = 14 cm

Answer: the dimensions of the rectangle are 14 cm x 4 cm.

b) The perimeter of a regular decagon is 50 cm. Find the length of each side of the decagon.

Perimeter of regular decagon = 10(side length) = 50 cm

Side length = 5 cm

Answer: the length of each side of the decagon is 5 cm

c) The perimeter of an equilateral triangle is 36 cm. Find the length of each side of the triangle.

Perimeter of equilateral triangle = 3(side length) = 36 cm

Side length = 12 cm

Answer: the length of each side of the triangle is 12 cm

d) The perimeter of a square is 68 cm. What is the length of each side of the square?

Perimeter of square = 4(side length) = 68 cm

Side length = 17 cm

Answer: the length of each side of the square is 17 cm

Exercise 4

Questions

a) The perimeter of a regular hexagon is 36 cm. Find the length of each side of the hexagon.

Perimeter of regular hexagon = 6(side length) = 36 cm

Side length = 6 cm

Answer: the length of each side of the hexagon is 6 cm

b) The perimeter of a rectangle is 28 cm. If the length is 3 cm more than twice the width, what are the dimensions of the rectangle?

Perimeter of rectangle = 2(length + width) = 28 cm

length = 3 cm + 2(width)

2(3 cm + 2(width) + width) = 28 cm

6(width) = 22 cm

width = 3.67 cm (rounded to two decimal places)

length = 3 cm + 2(3.67 cm) = 10.33 cm (rounded to two decimal places)

The dimensions of the rectangle are 10.33 cm x 3.67 cm.

c) The perimeter of a regular octagon is 64 cm. Find the length of each side of the octagon.

Perimeter of regular octagon = 8(side length) = 64 cm

Side length = 8 cm

d) The perimeter of a rectangle is 30 cm. If the length is 4 cm more than three times the width, what are the dimensions of the rectangle?

Perimeter of rectangle = 2(length + width) = 30 cm

length = 4 cm + 3(width)

2(4 cm + 3(width) + width) = 30 cm

8(width) = 22 cm

width = 2.75 cm (rounded to two decimal places)

length = 4 cm + 3(2.75 cm) = 12.25 cm (rounded to two decimal places)

Answer: the dimensions of the rectangle are 12.25 cm x 2.75 cm.

Exercise 5

Questions

a) The perimeter of a square is 44 cm. What is the length of each side of the square?

b) The perimeter of an isosceles triangle is 32 cm. If the base is 10 cm, what is the length of each of the other two sides?

c) The perimeter of a regular pentagon is 35 cm. Find the length of each side of the pentagon.

d) The perimeter of a rectangle is 34 cm. If the length is twice the width, what are the dimensions of the rectangle?

Answers

a) The perimeter of a square is 44 cm. What is the length of each side of the square?

Perimeter of square = 4(side length) = 44 cm

Side length = 11 cm

Answer: the length of each side of the square is 11 cm

b) The perimeter of an isosceles triangle is 32 cm. If the base is 10 cm, what is the length of each of the other two sides?

Perimeter of isosceles triangle = base + 2(sides) = 32 cm

Base = 10 cm 2(sides) = 32 cm - 10 cm = 22 cm sides = 11 cm

Answer: the length of each of the other two sides is 11 cm.

c) The perimeter of a regular pentagon is 35 cm. Find the length of each side of the pentagon.

Perimeter of regular pentagon = 5(side length) = 35 cm

Side length = 7 cm

Answer: the length of each side of the pentagon is 7 cm

d) The perimeter of a rectangle is 34 cm. If the length is twice the width, what are the dimensions of the rectangle?

Perimeter of rectangle = 2(length + width) = 34 cm length = 2(width)

2(2(width) + width) = 34 cm

6(width) = 34 cm

width = 5.67 cm (rounded to two decimal places)

length = 2(5.67 cm) = 11.34 cm (rounded to two decimal places)

Answer: the dimensions of the rectangle are 11.34 cm x 5.67 cm.

Part 6: Perimeter

Exercise 1

Questions

a) The perimeter of a square is 36 cm. What is the length of each side of the square?

b) The perimeter of an equilateral triangle is 42 cm. Find the length of each side of the triangle.

c) The perimeter of a rectangle is 20 cm. If the length is 3 cm less than twice the width, what are the dimensions of the rectangle?

d) The perimeter of a regular heptagon is 21 cm. Find the length of each side of the heptagon.

Answers

a) The perimeter of a square is 36 cm. What is the length of each side of the square?

Perimeter of square = 4(side length) = 36 cm

Side length = 9 cm

Answer: the length of each side of the square is 9 cm

b) The perimeter of an equilateral triangle is 42 cm. Find the length of each side of the triangle.

Perimeter of equilateral triangle = 3(side length) = 42 cm

Side length = 14 cm

Answer: the length of each side of the triangle is 14 cm

c) The perimeter of a rectangle is 20 cm. If the length is 3 cm less than twice the width, what are the dimensions of the rectangle?

Perimeter of rectangle = 2(length + width) = 20 cm

length = 2(width) - 3 cm 2((2(width) - 3 cm) + width) = 20 cm

6(width) = 26 cm

width = 4.33 cm (rounded to two decimal places)

length = 2(4.33 cm) - 3 cm = 5.67 cm (rounded to two decimal places)

Answer: the dimensions of the rectangle are 5.67 cm x 4.33 cm.

d) The perimeter of a regular heptagon is 21 cm. Find the length of each side of the heptagon.

Perimeter of regular heptagon = 7(side length) = 21 cm

Side length = 3 cm

Answer: the length of each side of the heptagon is 3 cm

Exercise 2

Questions

a) The perimeter of a parallelogram is 34 cm. If the length is twice the width, what are the dimensions of the parallelogram?

b) The perimeter of an isosceles triangle is 28 cm. If the base is 8 cm, what is the length of each of the other two sides?

c) The perimeter of a square is 48 cm. What is the length of each side of the square?

d) The perimeter of a rectangle is 24 cm. If the length is 3 cm more than twice the width, what are the dimensions of the rectangle?

e) The perimeter of a regular hexagon is 36 cm. Find the length of each side of the hexagon.

Answers

a) The perimeter of a parallelogram is 34 cm. If the length is twice the width, what are the dimensions of the parallelogram?

Perimeter of parallelogram = 2(length + width) = 34 cm

length = 2(width) 2(2(width) + width) = 34 cm

6(width) = 34 cm width = 5.67 cm (rounded to two decimal places)

length = 2(5.67 cm) = 11.34 cm (rounded to two decimal places)

Answer: the dimensions of the parallelogram are 11.34 cm x 5.67 cm.

b) The perimeter of an isosceles triangle is 28 cm. If the base is 8 cm, what is the length of each of the other two sides?

Perimeter of isosceles triangle = base + 2(sides) = 28 cm

Base = 8 cm

2(sides) = 28 cm - 8 cm = 20 cm

sides = 10 cm

Answer: the length of each of the other two sides is 10 cm.

c) The perimeter of a square is 48 cm. What is the length of each side of the square?

Perimeter of square = 4(side length) = 48 cm

Side length = 12 cm

Answer: the length of each side of the square is 12 cm

d) The perimeter of a rectangle is 24 cm. If the length is 3 cm more than twice the width, what are the dimensions of the rectangle?

Perimeter of rectangle = 2(length + width) = 24 cm

length = 2(width) + 3 cm

2((2(width) + 3 cm) + width) = 24 cm

6(width) + 6 cm = 24 cm

6(width) = 18 cm

width = 3 cm

length = 2(3 cm) + 3 cm = 9 cm

Answers: the dimensions of the rectangle are 9 cm x 3 cm.

e) The perimeter of a regular hexagon is 36 cm. Find the length of each side of the hexagon.

Perimeter of regular hexagon = 6(side length) = 36 cm

Side length = 6 cm

Answer: the length of each side of the hexagon is 6 cm

Exercise 3

Questions

a) The perimeter of a parallelogram is 28 cm. If the length is 5 cm more than the width, what are the dimensions of the parallelogram?

b) The perimeter of an isosceles triangle is 40 cm. If the base is 12 cm, what is the length of each of the other two sides?

Answers

a) The perimeter of a parallelogram is 28 cm. If the length is 5 cm more than the width, what are the dimensions of the parallelogram?

Perimeter of parallelogram = 2(length + width) = 28 cm

length = width + 5 cm

2((width + 5 cm) + width) = 28 cm

4(width) + 10 cm = 28 cm

4(width) = 18 cm

width = 4.5 cm

length = 4.5 cm + 5 cm = 9.5 cm

Answers: the dimensions of the parallelogram are 9.5 cm x 4.5 cm.

b) The perimeter of an isosceles triangle is 40 cm. If the base is 12 cm, what is the length of each of the other two sides?

Perimeter of isosceles triangle = base + 2(sides) = 40 cm

Base = 12 cm

2(sides) = 40 cm - 12 cm = 28 cm

sides = 14

Answer: the length of each of the other two sides is 14 cm

Conclusion

Thank you once again for purchasing this book. I hope it has helped you in your journey to understand the basics of perimeter.

Please, if you learnt something from this book, I would like you to leave a review. It'd be appreciated.

Thank you.